YOUR KNOWLEDGE HAS VALUE

- We will publish your bachelor's and
  master's thesis, essays and papers

- Your own eBook and book -
  sold worldwide in all relevant shops

- Earn money with each sale

Upload your text at www.GRIN.com
and publish for free

Bibliographic information published by the German National Library:

The German National Library lists this publication in the National Bibliography;
detailed bibliographic data are available on the Internet at http://dnb.dnb.de .

Imprint:

Copyright © 2011 GRIN Verlag, Open Publishing GmbH
Print and binding: Books on Demand GmbH, Norderstedt Germany
ISBN: 9783656073512

This book at GRIN:

http://www.grin.com/en/e-book/182962/phytoremediation-in-reed-plants-treat-
and-clean-up-polluted-environment

Ochan Stephen

Phytoremediation in reed plants treat and clean up polluted environment by petroleum produced water

GRIN Publishing

OCHAN STEPHEN

Phytoremediation in reed plants treat and clean up polluted environment by petroleum produced water.

A Final Thesis Presented to the Academic

Department Of the school of Science and Engineering in

Partial Fulfillment of the Requirements

For the Undergraduate Degree of Petroleum Engineering.

ATLANTIC INTERNATIONAL UNIVERSITY

HONOLULU, HAWAII June 10, 2011

Table of Contents

APPROVAL

Name: Ochan Stephen

Degree: Bachelor of Science in Petroleum Engineering

Title of Thesis: **Phytoremediation in reed plants treat and clean up polluted environment by petroleum produced water.**

Acknowledgements.

First of all, let me thank God almighty father for his eminent effort and love bestowed to me in realizing talent as a gift of the Holy Spirit.

Many gratitudes and thanks goes to my academic advisors **Dr. Randall Raus** and **Mr. Jorge Vasconcelos-Santillan** of AIU for their supervision and mentorship throughout my thesis working. Friendly coordination and skillful guidance under their supervision has been beneficial. They always respond to my attention whenever need be for the complete and successful evolution of this thesis.

I would like to thank Mr. Jason and Marks my immediate technical supervisors for their real time dedication of performing some laboratory tests and full feed back from a giant laboratory in UK during which they were also assigned to carry out similar tests in petroleum waste bioremediation, which I benefited a lot.

I would like to thank my parents and family members especially my Mum Rose Nadio Peter and Dad Simon Peter Lochi whom for a long never spent my time with them since I was away for the reason best known to them.

Lastly many thanks and appreciations are poured to (WNPOC) White Nile petroleum Operating Company Engineers and Technical personnel for permission given to me to carry out of this thesis research and for using their laboratory and process facilities for oil production which without them this work won't be fruitful now.

Abstract.

Phytoremediation is the use of plants and its associated microorganisms to achieve the conditions necessary to facilitate the breakdown of contaminants and clean-up of the polluted environment. Phytoremediation technology is viewed as the simplest way of handling variety of contaminants in many sectors of oil industry. The community of microorganisms in the rhizosphere has been shown to be involved in degradation of numerous contaminants, including pesticides, polynuclear aromatic hydrocarbons, petroleum compounds, volatile organic chemicals, and in organics. Also, plants can degrade contaminants during plant metabolic activities; for instance, 2,4,6-trinitrotoluene has been shown to be degraded by plant enzymes. Plants can use contaminants as nutrients; nitrate contamination of ground water can serve as a nitrogen source for plants. This involves the achieving condition of ground water, waste oil and produced water from oil facilities.

Phytoremediation is recommended because of its establishment at low -cost and with flexibility in wide aspect of soil environmental remediation. Practical field experiments and laboratory water quality test analysis carried out in an oil field in TharJath-south Sudan by using reed species called **Phragmites australis** has shown promising results. It is common perennial grasses that thrive best in wetlands and temperate tropical part of the world. Southern part of the Sudan being temperate and tropical region is suitable for this reed species to thrive well. This Thesis provides a real analytical test and report on the effective clean up of the polluted environment by the use of reed plants through phytoremediation process. Hence will encourage future studies on remediation of the contaminated soil.

Abbreviations /Acronyms

- COD   Chemical Oxygen demand
- TDS   Total Dissolved Solids
- BOD   Biological Oxygen Demand
- pH   Power of Hydrogen in water.
- CPF   Central Processing Facilities
- WNPOC   White Nile Petroleum operating Company
- BWPD   Barrels of Water Per Day
- EC   Electrical Conductivity
- BS&W   Base Sediment and Water
- PPM   Parts Per Million
- EPA   Exploration Production and Analysis.

Chapter 1. Introduction and background.

Phytoremediation natural process to clean up the polluted environment from pollutants or contaminants is the focal point in an emerging technology .Such pollutants or contaminants range from metals such as Zinc ions, sodium ions, chlorides, radionuclide, chemicals from oil treatment processes such as demulsifier and reverse demulsifiers. The selected plant species for this technology is reed (**Phragmites australis)**. It's a perennial plant that can thrive well in wetland with minimum sunlight as will be discussed in chapter 2. Petroleum wastes are documented to naturally degrade in natural wetland environments (**Wemple &Hendricks, 2000)**. The microbial community associated with the plant rhizosphere creates an environment conducive to degradation of many volatile organic compounds (**Schnoor** *et al.*, **1995; Pardue** *et al.*, **2000)**. Subsurface flow constructed wetlands (CWs) have been used to treat petroleum wastewaters (Knight *et al.*,1999).Field trials have shown successful removal of petroleum hydrocarbon contaminants by **in situ** remediation (**Cunnigham** *et al.* **2001**). Aromatic hydrocarbons, BTEX (benzene, toluene, ethyl benzene, trimethylbenzene) and MTBE (methyl tert-butyl ether) are groundwater contaminants at a former refinery site in Leuna (Germany).
The use of reeds for the treatment of sewage was first investigated in Germany by Seidel and Kickuth in the 1960's. Since then, about 500 reed bed treatment systems have been constructed in Western Europe since 1984. In general, the experiences gained in the years since, show that Biochemical Oxygen Demand removal (BOD) is 80 to 90%, with typical outlet concentrations of 20 ppm.

Phytoremediation concept is based on the well-known ability of plants and their associated rhizospheres to concentrate and/or degrade highly dilute contaminants. Critical components of the rhizosphere, in addition to a variety of free-living microorganisms, include root exudates. These complex root secretions, which "feed" the microorganisms by providing carbohydrates, also contain natural chelating agents (citric, acetic, and other organic acids) that make the ions of both nutrients and contaminants more mobile in the soil. Root exudates may also include enzymes, such as nitroreductase dehalogenases, and laccases. These enzymes have important natural functions, but they may also degrade organic contaminants that contain nitro groups (e.g., TNT, other explosives) or halogenated compounds (e.g., chlorinated hydrocarbons, many pesticides). Phytoremediation technology emerge in recent years a flexible approach for cheaper waste minimization and management in oil and gas industry. Many researchers such as Applied Natural Science conducted a green house experiment on Zinc uptake by hybrid poplar (populus sp) was initiated in 1995. This indicated that high levels of zinc (4,200 μg/g [ppm]) in leaves of hybrid poplar growing as a cleanup system at a site with zinc contamination in the root zone of some of the trees. However, the focus of this thesis is to prove that phytoremediation is effective by use of the species selected by performing real water quality test analysis on the concentration of Na+ ions, Cl_ ions (COD) chemical oxygen demand, (BOD) Biological Oxygen demand, pH, Temp, (DO) Dissolve Oxygen,Disolve Oil. This plant biological approach will enable free disposal of this treated water in an environmentally friendly manner. For soils contaminated with toxic organics, the approach is similar, but the plant may take up or assist in the degradation of the organic (**Schnoor et al. 1995).**

Research Hypothesis

This thesis on phytoremediation discusses the adoption of this technology in groundwater treatment and produced water from oil wells and processing facilities which involves practical analysis in field laboratory trail tests of daily sample reports and data parameters presenting actual effectiveness of the technology in TharJath oil field in south Sudan.
It also investigates the adaptation of reed plant species for the sequestration or degradation of specific contaminants by evaluating the water quality parameters such as the pH, Biological oxygen demand (BOD), Dissolve oxygen, chemical oxygen demand (COD) Electrical conductivity (EC) of the solution total dissolved solids (TDS) and botanical leave monitoring results periodically at intervals.
The thesis also aims at integrating the current research by OCEANSES for global environmental solutions carried out in Sudan oil fields. The application of reed beds are useful for the clean up of waste in areas such as, Oil field waste water, Chemical industry effluent, Composting leachate ,Pesticide contaminated water, Run off from Petrol Stations, Contaminated groundwater, Fire Training water, Domestic sewage/grey water.

This examines the effectiveness and efficiency of the plant's natural process in break-down of such organic compounds from petroleum production. Therefore, when the selected species is grown, adapted plants in contaminated substrates, the root system functions as a highly dispersed, fibrous uptake system. Contaminants over a large range of concentrations are taken up along with the water and degraded, metabolized, and/or sequestered in the plant body, while evapotranspiration from aerial parts maximizes the movement of soil solution or wastewater through the plant.
 Through the process of bioaccumulation, contaminants can be concentrated thousands of times higher in the plant than in the soil or wastewater. The contaminated plant biomass can be digested or ashed to reduce its volume (_95%), and the resulting small volume of material can be processed as an "ore" to recover the contaminant (e.g., valuable heavy metals, radionuclides).

If recycling the metal is not economically feasible, the relatively small amount of ash (compared to the original biomass or the extremely large volume of contaminated soil) can be disposed of in an appropriate manner.

The questions asked here are, how do reeds by phytoremediation process help in cleaning up the contaminated waste water by oil, chemicals such as pesticides and other metals? What stage of growth in reeds does phytoremediation process occur? And for how long does this process take? What are the long term benefits of the treated waste water to the ecology and ecosystem? Howe ever, many considerations such as the amount of pollutants or harmful chemical present should be understood, size and depth of the polluted area, type of soil and conditions present. These factors will necessitate fruitful and promising results from the process.

Study area

Sudan is the Africa's largest country with an estimate of 40 million people. It is one of the worst hit countries by war for two decades which finally signed the comprehensive peace agreement (CPA) on 9th, January, 2005 in Kenya that brought war to a halt. It is one of the third richest countries with oil reserves in Africa, research proved most of the oil is found in the southern part of the country. Sudan is currently producing over 450,000 (BOPD) barrels of oil per day to sustain its economy dependence of up to 98% gross income. However, with the environmental impacts and management of byproducts from oil and effluent discharged in a controlled manner, the oil fields adapted bioremediation project to help minimized millions of barrels of produced water disposed into the environment. In Bentiu,Unity State of south Sudan, depending on the volume of produced waste water from the entire (CPF) Central processing facilities in Tharjath and Mala (DGFF) degassing field facilities of petroleum fields, and daily total produced water which is between 12,000 -18,000 (BWPD)Barrels of water per day . This waste water passes through water treatment in the CPF facilities such as the (CPI) corrugated plate interceptor in which it is geometrically designed and installed with skimmers that help separate oil from the produced water from (FWKO) Free Water Knock Out tank to achieve a maximum of 2000ppm before it flows to (IGF) Induced Gas Floatation. IGF principles through the help of motors that drive the parallel skimmers to induced the treated water from CPI by forming a foam like substance that collects the oil at the surface which is then recycled into the process by IGF sump pit pump. The 100 ppm water flow at the bottom through a pipe into the retention and evaporation ponds for further gravitational separation which later pumped back the emulsion into the process. These ponds are three in number, Retention pond 1, 2, 3 and Evaporation pond.

Daily samples are collected for laboratory water quality test analysis before being pumped out to reed beds for phytoremediation process. The bioremediation project has 17 lagoons that retain the water in which the reeds grow to their maximum span. Since reeds begin phytoremediation process at early stage and continue throughout their lives, daily water analyses from each lagoon inlet and outlet and other weir, valves are taken for laboratory test. Lagoons are pond-like bodies of water or basins designed to receive, hold, and treat wastewater for a predetermined period of time. If necessary, they are lined with material, such as clay or an artificial liner, to prevent leaks to the groundwater below.

The general area for carrying out this project is about 23,000 hectors of land which holds over 120000-180000 (BOWPD) and it's the second largest bioremediation project in the world managed by (WNPOC) White Nile Petroleum Operating Company in partnership with PETRONAS Malaysia in block 5A in Tharjath unity state. An up-to-date laboratory for this project is installed at the site for sample tests and water quality analysis and generation of reports to be sent to UK giant laboratory for further scientific analysis. Oceans global environmental solutions established this project to enable Sudan handle it's waste from oil production from hampering the environment by disposal or any oil spill. Oceans integrate a scientific approach to solve number of wastes contributing to pollution worldwide especially in oil producing countries such as USA, China, South Africa and all other industrial waste from factories. Sewage system in the UK is being handled by bioremediation and achieved by reeds plants through phytoremediation process.

Reeds are planted in lagoons; Lagoons are designed by qualified professionals who have had experience with them. Certain site-related factors, such as the location of the water table and the composition of the soil, always must be considered when designing lagoon systems. Ideally, lagoons should be constructed in areas with clay or other soils that won't allow the wastewater to quickly percolate down through the lagoon bottom to the groundwater. Otherwise, lagoons must be artificially lined with clay, bentonite, plastic, rubber, concrete, or other materials to prevent groundwater pollution. Special linings usually increase system costs.

The structural design of Lagoon in TharJath oil field.

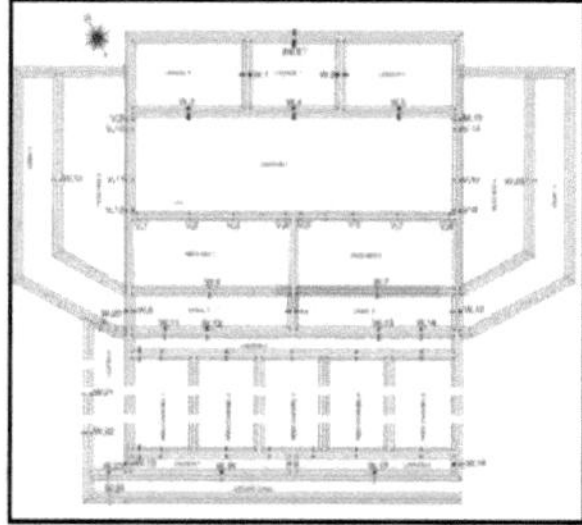

Photo shows current lagoon photo for bioremediation established by OCEANS in TharJath oil field.

Reed beds establishment

Reeds within both beds were showing signs of healthy growth. Runners have been observed throughout the beds, which over time will increase percentage cover. Approximately 50% of the reeds within both reed beds were reproducing.

Irrigation of reed beds is done on continuinoius basis and all areas showed good rates of drainage that can enhance ample growth.

At some point, approximately 5% of Reed Beds 1 and 2 in appendix B show poor signs of growth due to a prolonged dry spell in April. During the rainy season, new growth has always been observed in these areas as shown in Reed Beds 3 and 4 in appendix B.

Reeds have reached a height of between 1.2 and 1.8m. Ample rain to the reed beds aid this growth and development.

Parameters measured and Frequency of the test analysis.

Parameters Measured	Frequency
Oil in Water (Dissolved and Free)	Daily for the inlet and outlet, Once per week for the lagoons and reed beds.
Dissolved Oxygen	Once per Week
Biological Oxygen Demand	Once Per Month
Electrical Conductivity	Once per Week
Temperature	Once per Week
Chemical Oxygen Demand	Once per Week for the Inlet, Once per 2 weeks in other locations)
pH Value	Once per Week
Total Dissolved Solids	Once per Week
Toxicological	Once per Three (3) Months
Botanical (Biomass development, leaf analysis and photosynthesis measurement)	Once per Month

Chapter 2: Water quality and laboratory testing for Chemical oxygen demand (COD) sample test.

Chemical Oxygen Demand (COD) is a measure of the total concentration of dissolved oxygen required to completely oxidise a dissolved material (either inorganic, e.g. nitrite/ammonia; or organic, e.g. dissolved hydrocarbons). As the COD increases generally water quality decreases. Below are instruments use for data analysis.

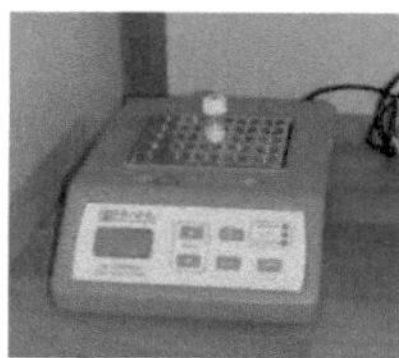

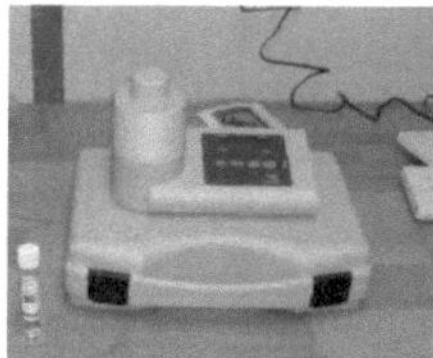

COD Heater Block **COD analyser.**

Figure1.4 shows COD analysers.

Procedures used;

Samples for COD and pH test were taken from inlet and outlet at temperature of 23 C° Celsius. The samples from both points were put in the conical flask and zero calibration was preformed by the COD analyzer and the results were recorded for inlet and outlet with 79 and 68 respectively. 2ml of sample were put into the conical flask, and few drops of sulphiric acid in each   W1-14 were shaken and temperature was recorded. The sampled solution was allowed to settle for 2- 4hrs. COD meter reactor/block was calibrated to zero tests and the samples were inserted in each tube for final test. The results were generated and plotted on the graph as shown below for further analysis. The graph shows Chemical Oxygen Demand at the Inlet and Outlet

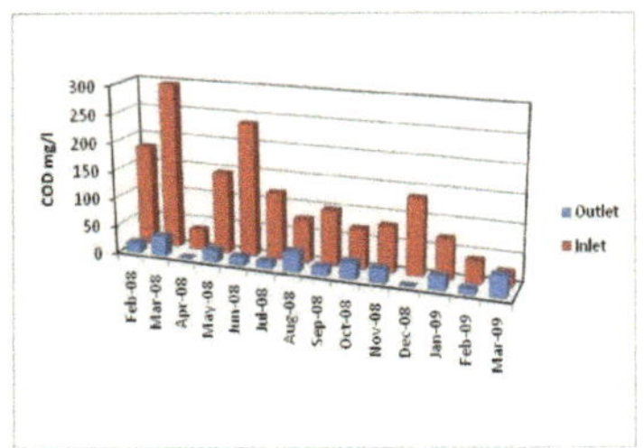

Results and discussion.

High COD readings at the inlet represent the dissolved oxygen required to oxidise both dissolved and free oil in the produced water. Chemical oxygen demand at the outlet represents the oxidisable materials present within outlet waters.

The acceptable presentation and integration for water quality analysis indicates the effluent treated by phytoremediation process is effective since measurement of oxygen concentration that dissolved in the water in order to completely oxidize the inorganic materials which was high at the inlet and low at the outlet due to microbial conversion in reed roots.

Note;
As the COD increases generally water quality decreases.

pH test

The pH test for each sample as graphically shown indicates gradual increase from acidity to alkalinity from the inlet to the outlet in every lagoon during the months of May June, July, August, Oct, November through Dec 2009 and January 2010-March 2010. It should be noted that the pH is between the Sudanese limits, while higher pH at the outlet indicate salts and minerals. The simple explanation shows that as the acidity and alkalinity is reached as result of polluted water that has been utilized by the natural process of phytoremediation through evotranspiration, transpiration from leaves used for plants growth. This means the bacteria in the roots such as the rihzophere can thrive well to play their role of metabolically convert the acidic water inside the reed plants through the entire growth by nitrification making nitrogen available for the plants growth. Generally the optimum pH value for a freshwater system is between 6.5 – 8. Below is the pH meter representation by colours.

pH representation.

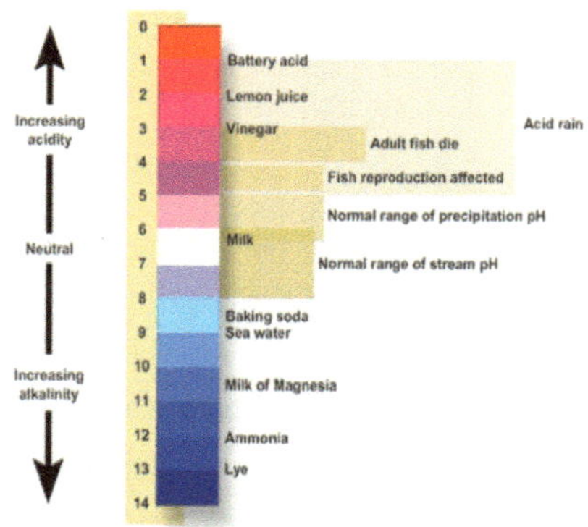

Sudanese maximum discharge limit - pH 10

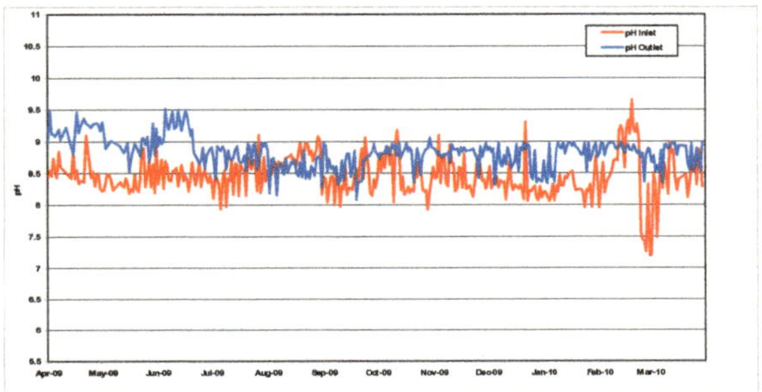

Sudanese minimum discharge limit - pH 6

Biological Oxygen Demand (BOD) weekly test.

Biological Oxygen Demand (BOD) is a measure of the total dissolved oxygen required to oxidise all biologically available material (although not including biological nitrification). This is in contrast to COD, which is a measure of all oxidisable material. BOD is a measure of the oxygen used by micro organisms as these materials are decomposed.

The BOD test measures the oxygen consumed in (mg/l) over 5days at 23 degrees C.

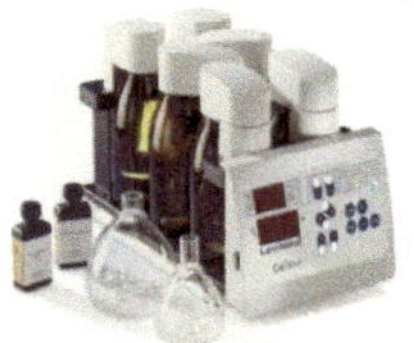

BOD sampler.

Methods and materials used;

The samples collected for BOD test were tested for pH which must be between 7-7.5 for neutrality and enable the microorganisms that live at that pH. When reading the pH the sample was acidic i.e. pH less than 7, Hcl acid solution of 1% concentrations was added to neutralize the sample for standard test. Although certain microbes live at pH as low as 1 which is acidic soil.

2ml of pillow were added together with BOD nutrient buffer pillow into the solution in the bottle. 10 drops of nitrification inhibitor plus 4 drops of potassium hydroxide solution. The samples were inserted into the BOD meter after the actual volume in ml was determined by the meter to be 428 ml. the time taken to start the programmed BOD meter was recorded and the samples were left for a period of 5 days.

Results and discussion.

The results show that the biological oxygen required for the micro organisms that live in the treated waste water by phytoremediation shows oxygen is present enough for oxidation of waste and organic compounds. This further explained that, enhanced biodegradation of pollutants at the plant roots in which the compounds exuded by the roots increase microbial biodegradation activity. This is achieved by uptake of plant roots that can enhance biodegradation by drawing pollutants towards the root zone of rhizophere and oxygen presence increased rhizophereremediation. The reaction of the biological oxidation occurs as below.

Oxidizable material +bacteria+nutrient+ $O_2 \rightarrow CO_2 + H_2O$ + Oxidized in organics such as NO_3 or SO_4. Oxygen consumption by reducing chemicals such as sulfides and nitrites is typified as follows.

$S + 2O_2 \rightarrow SO_4$

$NO_2 + 1/2O_2 \rightarrow NO_3$.

Since such produced water contains living bacteria together with the nutrients such as the above reactions, the 5-day BOD measures the amount of oxygen consumed by biological oxidation of waste contaminants in a 5-dya period. This measure the total amount of oxygen consumed when the biochemical reaction is allowed to proceed to completion is called the Ultimate BOD. This process is too time consuming, so the 5-day BOD has almost universally been adopted as a measure of relative pollution effect.

Water sources showing the standard units for various samples carried out in each respective sites;

<u>Typical BOD Values as an example:</u>
Pristine River - < 1m g/l
Polluted River- 2-8mg/l

Na+ ions test from inlet and outlet (daily)

The temperature for each sample form inlet and outlet were taken, an addition of Na+ ions standard solution of 1000ppm in 500 ml conical flask was set for zero calibration to this standard solution. The inlet and the outlet solution were then measured by dipping the meter probe and agitated. The inlet indicates 698 Na+ ions while the outlet indicates 740 Na+ ions. The below graph shows monthly concentration of ions in water.

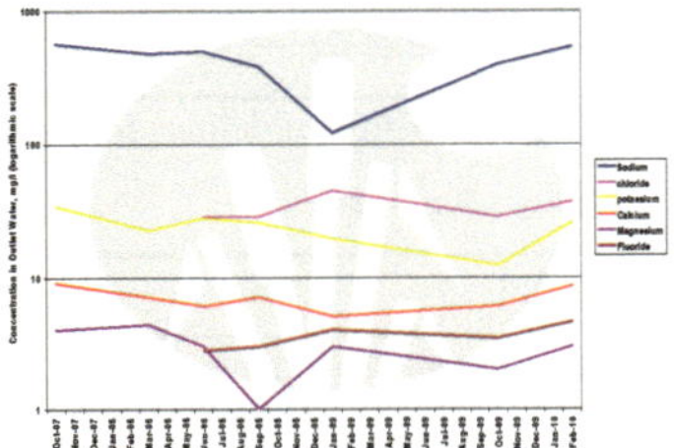

Chapter 3. Data analysis and presentation of water.

Daily sampling and analysis of water is being carried out at locations where water is present. The provisions of weekly results for the months are provided in Appendix A. Dissolved oil analysis presents promising results; the readings show a healthy biodegradation of hydrocarbons through the system as shown in Appendix A.

- Parameter data is being collected at sites where produced water is present.
- Dissolved oil analysis the readings show the biodegradation of hydrocarbons through the system. The results illustrated appear to differ on occasion. This is due to the gradual increase of produced water being sent to the system, periods of heavy rain and the inlet walkway that has been fitted.

The electrical conductivity and Total dissolve Solid (EC & TDS) is slightly elevated in some areas due to the stagnation of the water and the effect of evaporation which is concentrating any dissolved salts.

Water samples were collected from the inlet and Weir 5 in April for trace element analysis. Preliminary results indicate that water quality is within expected parameters.

Table below shows showing comparison of water quality parameters with Sudanese standard limits.

Parameter	Sudanese Standard Limit
Oil and Grease	3 mg/l
Dissolved oxygen	Above 2mg/l (~20% at outlet temperature)
Electrical Conductivity	0-3000 µS (FAO guidelines for Irrigation water quality standards)
BOD	35 mg/l
Temperature	No more than 2°C above ambient temperature
COD	80 mg/l
pH	Between pH 6 & 10

Monitoring and mentoring.

Monitoring of the system gives a summary of what is occurring over time biologically, chemically and physically. The development of the entire growth of the reed and the effectiveness in producing exudates play major role in produced water. Scientific results from laboratory prove toxic free water can be dispose to environment. The wildlife and other agricultural investment activities can be carried upon by utilisation of the treated water; Such as planting of trees and cultivation of cash crops, fishing as fish thrive well.

Parameter Measurements; as described in chapter 2 the parameters which were monitored regularly (on a day by day or week by week basis) on the system included:

- Dissolved and Free Oil (free oil measured in production lab)
- Temperature
- Dissolved Oxygen
- Total Dissolved Solids
- pH
- Electrical Conductivity
- Chemical Oxygen Demand and Biological Oxygen Demand

The Central processing facilities of oil production send the produced water every week and this was included in the weekly and monthly reports for the period of 6 months. The volume of water generated from the process is pumped and running hours of the pumps with each capacity in volumes are tabled as below. This will enhance the proper monitoring and mentoring of bioremediation for phytoremediation process.

Dissolved Oxygen

Dissolved oxygen (DO) analysis measures the amount of gaseous oxygen dissolved in an aqueous solution. Oxygen gets into water by aeration (rapid movement), diffusion from the surrounding air and as a waste product of photosynthesis.

Dissolved Oxygen is a measure of good water quality although levels should not exceed 110% as this can be harmful to aquatic life. It is considered that if dissolved oxygen is less than 5mg/l then aquatic life is put under stress and if oxygen levels read below 1-2mg/l this can result in large fish kills

Dissolve Oxygen meter.

Results of low oxygen.

This explains that the time of day can also have a effect on DO measurements. Later on during the day there is normally more DO due to photosynthesis of reeds.
Other factors effecting DO, including water temperature, atmospheric pressure and altitude, and salinity of the solution

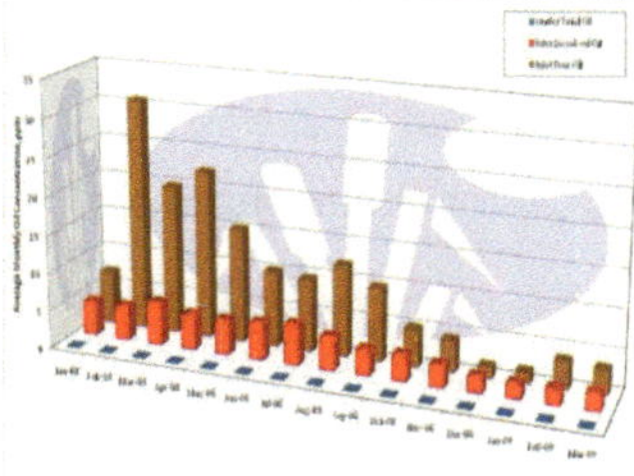

The graph above shows free and dissolved oil at the inlet compared to the oil at the outlet.

Fluctuations in dissolved and free oil concentration may be attributable to changes in the production processes, or the bringing online of new wells with different oil characteristics. There can also be a seasonal effect as rainwater can dilute the samples – so lower oil content in the wet season.

Oil in Water

Oil in water is the measurement of dissolved and free hydrocarbon content within an aqueous solution.

It is measured because high concentrations can cause serious damage to the environment, such as habitat loss and the pollution of groundwater and is the main reason for why we use Bioremediation.

Dissolved oil is measured using the fluo-imager in the laboratory and free oil is measured in the production laboratory.

Inlet

Outlet

As you can see from the photos the difference, visually in water quality. In the first photo you can see the FREE oil whereas at the outlet this cannot be seen.

Shows pump running hours for water supply in volume

Date	Running time (hrs)	Pump No.	Approximate volume (M³)
Mon 28th	7.5	1	3,975
Tue 29th	-	-	-
Wed 30th	-	-	-
Thurs 1st	-	-	-
Fri 02nd	4.5	1	2,385
Sat 03rd	-	-	-
Sun 04th	6.6	1	3,300
Total Volume			9,660

The monitoring of water entering the lagoons are recorded and plotted as below.

Graph showing Produced Water Volumes Compared to Previous Years

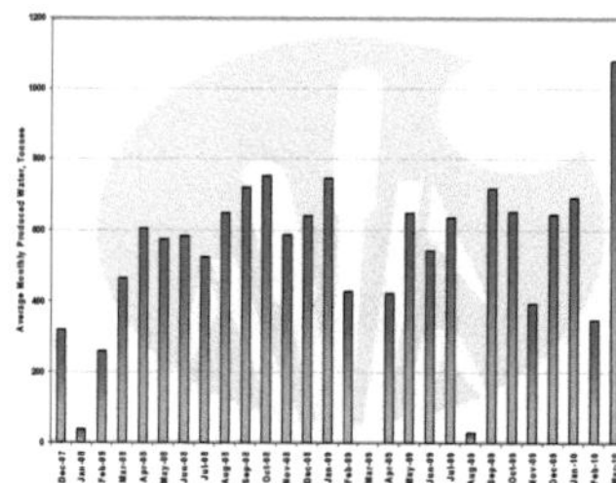

Biomass Reports

At the end of each month a Biomass Report is conducted to give a representative of reed health and growth.

In each red bed (of the two working) there are 3 areas of 1m2 whereby the survey was conducted. These areas are marked out with wooden stakes so that each survey is obtained from the same areas every time.

Once the data was compiled, then comparison was done for previous months to show any signs of reed growth and health.

Growing reeds

Biomass monitoring continued on monthly basis to assess the performance of the reeds and records of entire parts of the plant are tabled as below.

Biomass monitoring in each area

Category	Area 1	Area 2	Area 3
Stem Number	22	27	30
Shoot Number	6	5	1
Runners Present	4	0	1
Ave. shoots - runners	1	0	0
Ave. stem thickness (mm)	7.25	5.8	6.3
Ave. Stem height (mm)	1050	775	1015
Ave. leaves/stem	11.75	13	13
Leaf area at 1M (Or nearest leaf to height)	412.5	303.75	1650

Chapter 4. Problem solving and presentations.

Further scientific research in reeds is still significant for the near future bioremediation of petroleum wastes in industries and other drilling activities of oil and gas. The focal point is mainly on safe handling of the residues or debris of the fluids used in drilling activities and petroleum production.

Evidences have shown reliable way of handling waste and chemicals on the planet earth through repetitive research such as the use of reeds.

It's not clear how dangerous will the selected species is to animals that feed on them or evaporation during plants transpirations to the atmosphere.

However, research indicates that nothing toxic is release in the environment due to effective metabolic activities of the micro-organisms found in the roots of the plants as exudates.

The healthy growth of the reeds established in bioremediation lagoons indicate significant conversion of the toxic chemicals in produced water for optimum growth.

The sample of water from varying water inlets and outlets indicate positive results by constant reduction of concentration of chemicals tested in laboratory due to effective reeds process.

Conclusions of Research and significance of future plans.

The basis of this thesis forms the most cheap and flexible way of handling petroleum waste by using reeds. The practical water sample analyses in various sites indicate promising results depicted from the contaminants contents in ground water.

The biomass and optimum growth of the reeds in beds reduction of contaminants at all levels showed effectiveness of the species in carrying out independent metabolic process.

The volume of treated water by the plants can be used for other agricultural activities such as cash crop cultivation etc.

Presence of aquatic life in the lagoons also indicates clean water treated and hence viable for life therefore, phytoremediation process is the real time solution in petroleum produced water.

To help us understand the petroleum waste disposal in safe manner, future plans should include the followings:

- Field demonstration of heavy metals removal and toxicity of the pollutants by the selected species including the biomass results.

- Conduct high resolution microanalysis of hyperaccumulator species (scanning or transmission electron microscopy with x-ray dispersive spectroscopy) to find the discrete sites of metals sequestration and bioaccumulation in specific plants organs, tissue, and organelles.

- Evaluate procedures for the disposal, processing, and volume reduction metals contaminated plant biomass.
- Conduct studies of roots and other plants biomass decomposition in soils to understand kinetics and cycling of contaminants.

- Evaluate the removal and sequestration of radionuclides from soil using plant-based clean up system.

- Extend investigations of the most promising lines of current research on plant-based clean up systems:

  - Phytoremediation using trees and large, robust, herbaceous plants

  -Degradation of halogenated organics.

  -Bioreactors of waste water clean up and volume reduction.

List of references.

1. American Petroleum Institute, Washington, DC (1991).

2. Stilwell, C.T., "Area Waste-Management Plans for Drilling and Production Operations," Journal of Petroleum Technology 67-71 (Jan. 1991).

3. "Guidance to Hazardous Waste Generators on the Elements of a Waste Minimization Program," 58 Federal Register 31114-31120 (May 28, 1993).

4. White Nile Petroleum Operating Company HSE management 2010.

5. Hinchman, R. and C. Negri. 1994a. The Grass Can Be Cleaner on the Other Side of the Fence. 1994. *logos*, Argonne National Laboratory, 12(2):8-11(Summer 1994).

6. Hinchman, R. and C. Negri. 1994b. A Greener Way to Handle 9. Wastewater & Avoid High Disposal Costs. *Econ.* 9(9):14-15, 39. (Sept. 1994).

7. Negri, M.C. and R.R. Hinchman. 1996. Plants That Remove Contaminants From the Environment. *Laboratory Medicine* 27(1): 36-40.

8. Negri, M.C., R.R. Hinchman, and E.G. Gatliff. 1996. 11.Phytoremediation: Using Green Plants to Clean Up Contaminated Soil, Groundwater, and Wastewater. *In*, Proceedings, International Topical Meeting on Nuclear and Hazardous Waste Management, Spectrum 96. Seattle, WA, August 1996.

9. American Nuclear Society.

10. Nyer, E.K. and E.G. Gatliff. 1996. Phytoremediation. *Ground Water Monitoring and Remediation* 16(1): 58-62.

APPENDIX A: Analytical Data

Samples are collected between system compartments. Free oil analysis is provided by WNPOC production department.

Table key.

 W= weir gate
V= valve

Dissolved Oil (DO)

Date	Sample Locations								
	Inlet	W1	W2	W3	W4	W5	W6	W7	V1
25th	6.11	7.14	6.45	3.77	5.28	4.42	-	-	2.08
26th	9.66	7.45	-	3.77	6.21	4.26	0.97	0.80	1.82
27th	7.79	6.68	6.43	3.54	5.53	3.75	0.51	0.92	1.89
28th	7.50	6.75	6.80	3.86	5.89	3.85	1.03	1.16	2.24
29th	6.59	6.11	5.95	3.64	5.47	3.59	1.00	1.00	2.10
30th	9.92	6.94	7.19	3.96	1.07	3.42	1.18	1.16	2.07
31st	11.04	-	-	-	-	-	1.72	-	-

pH

Date	Sample Locations										
	Inlet	W1	W2	W3	W4	W5	W6	W7	V1	V5	V10
1st	-	-	-	-	-	-	-	-	-	-	-
2nd	-	-	-	-	-	-	-	-	-	-	-
3rd	9.07	9.18	9.20	9.33	9.22	9.42	-	-	9.74	-	-
4th	9.12	9.20	9.18	9.28	9.26	9.22	-	-	9.56	-	-
5th	-	-	-	-	-	-	-	-	-	-	-
6th	-	-	-	-	-	-	-	-	-	-	-
7th	9.14	9.20	9.24	9.42	9.25	9.33	-	-	9.48	-	-
8th	9.10	9.18	9.25	9.36	9.22	9.40	-	-	9.44	-	-
9th	-	-	-	-	-	-	-	-	-	-	-
10th	9.14	9.26	9.30	9.46	9.28	9.38	-	-	9.42	-	-
11th	9.13	9.26	9.32	9.44	9.30	9.34	-	-	9.46	-	-
12th	9.10	9.20	9.37	9.44	9.56	9.52	-	-	9.53	9.54	-
13th	-	-	-	-	-	-	-	-	-	-	-
14th	9.12	9.29	9.33	9.48	9.33	9.42	-	-	9.48	9.54	-
15th	-	-	-	-	-	-	-	-	-	-	-
16th	9.13	9.27	9.35	9.47	9.30	9.41	-	-	9.46	9.58	-
17th	9.10	9.23	9.24	9.59	9.24	9.38	-	-	-	-	9.70
18th	-	-	-	-	-	-	-	-	-	-	-
19th	-	-	-	-	-	-	-	-	-	-	-
20th	-	-	-	-	-	-	-	-	-	-	-
21st	-	-	-	-	-	-	-	-	-	-	-
22nd	-	-	-	-	-	-	-	-	-	-	-
23rd	9.30	9.31	9.31	9.48	9.29	9.42	-	-	-	-	-
24th	9.19	9.37	9.35	9.49	9.39	9.46	-	-	9.19	-	-
25th	9.17	9.34	9.28	9.42	9.28	9.42	-	-	9.49	-	-
26th	9.15	9.28	-	9.49	9.28	9.39	9.86	9.79	9.45	-	-
27th	9.14	9.35	9.35	9.51	9.34	9.49	9.86	9.88	9.48	-	-
28th	9.17	9.33	9.26	9.40	9.34	9.46	9.86	9.81	9.50	-	-
29th	9.17	9.34	8.93	9.40	8.85	8.94	9.34	9.81	9.48	-	-
30th	8.61	8.84	8.82	8.99	8.84	8.88	9.28	9.30	9.05	-	-
31st	8.52	-	-	-	-	-	9.21	-	-	-	-

Temperature (°C)

Date	Sample Locations										
	Inlet	W1	W2	W3	W4	W5	W6	W7	V1	V5	V10
1st	-	-	-	-	-	-	-	-	-	-	-
2nd	-	-	-	-	-	-	-	-	-	-	-
3rd	31.0	30.2	31.1	29.5	30.3	30.8	-	-	33.6	-	-
4th	26.7	26.8	26.7	27.2	27.0	26.8	-	-	29.4	-	-
5th	-	-	-	-	-	-	-	-	-	-	-
6th	-	-	-	-	-	-	-	-	-	-	-
7th	30.1	29.2	28.7	29.0	29.1	28.8	-	-	30.6	-	-
8th	28.6	29.2	28.7	28.6	27.7	28.1	-	-	29.2	-	-
9th	-	-	-	-	-	-	-	-	-	-	-
10th	29.5	28.7	28.2	27.1	28.4	28.1	-	-	31.8	-	-
11th	27.6	27.9	28.0	28.1	27.6	27.8	-	-	-	-	-
12th	30.5	30.8	29.5	29.1	29.6	28.7	-	-	33.6	32.9	-
13th	-	-	-	-	-	-	-	-	-	-	-
14th	25.7	26.9	26.8	27.1	27.0	27.1	-	-	29.6	29.4	-
15th	-	-	-	-	-	-	-	-	-	-	-
16th	30.7	31.4	29.1	29.1	29.8	29.8	-	-	32.4	30.8	-
17th	30.5	30.4	29.8	29.7	29.8	28.8	-	-	-	-	29.5
18th	-	-	-	-	-	-	-	-	-	-	-
19th	-	-	-	-	-	-	-	-	-	-	-
20th	-	-	-	-	-	-	-	-	-	-	-
21st	-	-	-	-	-	-	-	-	-	-	-
22nd	-	-	-	-	-	-	-	-	-	-	-
23rd	27.9	27.4	28.1	28.4	28.0	28.3	-	-	-	-	-
24th	26.1	25.8	26.8	26.3	26.4	26.5	-	-	24.1	-	-
25th	29.4	29.6	30.2	30.0	30.6	29.7	-	-	33.8	-	-
26th	28.6	27.7	-	27.7	28.4	28.6	28.0	28.0	28.9	-	-
27th	28.6	27.3	26.8	27.1	27.5	27.2	26.9	26.6	27.5	-	-
28th	27.6	26.9	26.7	26.7	27.3	27.5	26.8	26.3	27.2	-	-
29th	29.8	29.5	29.6	30.4	28.6	30.2	30.8	32.5	28.1	-	-
30th	28.7	27.3	27.9	28.0	28.0	28.2	28.4	28.4	27.7	-	-
31st	-	-	-	-	-	-	-	-	-	-	-

Electrical Conductivity (mS/cm) (EC).

Date	Sample Locations										
	Inlet	W1	W2	W3	W4	W5	W6	W7	V1	V5	V10
1st	-	-	-	-	-	-	-	-	-	-	-
2nd	-	-	-	-	-	-	-	-	-	-	-
3rd	3.30	3.59	3.67	4.28	3.65	5.33	-	-	9.15	-	-
4th	3.32	3.54	3.66	4.24	3.58	4.86	-	-	7.98	-	-
5th	-	-	-	-	-	-	-	-	-	-	-
6th	-	-	-	-	-	-	-	-	-	-	-
7th	3.34	3.50	3.70	4.42	3.48	4.22	-	-	4.74	-	-
8th	3.22	3.52	3.58	4.26	3.66	4.33	-	-	4.78	-	-
9th	-	-	-	-	-	-	-	-	-	-	-
10th	3.13	3.55	3.75	4.95	3.61	4.24	-	-	4.68	-	-
11th	3.18	3.48	3.72	4.81	3.77	4.18	-	-	4.72	-	-
12th	3.19	3.66	3.54	4.39	3.71	4.95	-	-	4.98	5.16	-
13th	-	-	-	-	-	-	-	-	-	-	-
14th	3.14	3.61	3.62	4.81	3.65	4.12	-	-	5.22	5.51	-
15th	-	-	-	-	-	-	-	-	-	-	-
16th	3.19	3.65	3.89	4.81	3.60	4.28	-	-	5.48	4.88	-
17th	3.15	3.50	3.41	4.80	3.46	4.14	-	-	-	-	4.60
18th	-	-	-	-	-	-	-	-	-	-	-
19th	-	-	-	-	-	-	-	-	-	-	-
20th	-	-	-	-	-	-	-	-	-	-	-
21st	-	-	-	-	-	-	-	-	-	-	-
22nd	-	-	-	-	-	-	-	-	-	-	-
23rd	3.60	3.47	3.52	4.61	3.49	4.01	-	-	-	-	-
24th	3.13	3.49	3.54	4.55	3.52	3.52	-	-	2.92	-	-
25th	3.31	3.49	3.42	4.56	3.49	4.02	-	-	4.91	-	-
26th	3.20	3.45	-	4.53	3.47	4.05	5.95	5.06	4.58	-	-
27th	3.19	3.45	3.71	4.49	3.45	4.00	5.94	4.93	5.01	-	-
28th	3.20	3.47	3.49	4.43	3.47	3.99	6.03	5.04	4.48	-	-
29th	3.26	3.58	3.53	4.50	3.56	4.10	6.17	5.24	4.54	-	-
30th	3.18	3.56	4.48	3.51	3.51	4.09	8.13	5.27	2.25	-	-
31st	3.12	-	-	-	-	-	5.88	-	-	-	-

Total Dissolved Solids(TDS) (ppt)

Date	Sample Locations										
	Inlet	W1	W2	W3	W4	W5	W6	W7	V1	V5	V10
1st	-	-	-	-	-	-	-	-	-	-	-
2nd	-	-	-	-	-	-	-	-	-	-	-
3rd	1.64	1.80	1.85	2.15	1.84	2.67	-	-	4.58	-	-
4th	1.66	1.78	1.86	2.12	1.80	2.58	-	-	3.68	-	-
5th	-	-	-	-	-	-	-	-	-	-	-
6th	-	-	-	-	-	-	-	-	-	-	-
7th	1.58	1.80	1.84	2.11	1.82	2.34	-	-	2.36	-	-
8th	1.62	1.76	1.80	2.08	1.81	2.22	-	-	2.42	-	-
9th	-	-	-	-	-	-	-	-	-	-	-
10th	1.56	1.77	1.88	2.48	1.80	2.12	-	-	2.37	-	-
11th	1.58	1.79	1.81	2.46	1.82	2.08	-	-	2.40	-	-
12th	1.60	1.83	1.97	2.19	1.84	2.46	-	-	2.47	2.57	-
13th	-	-	-	-	-	-	-	-	-	-	-
14th	1.59	1.81	1.81	2.45	1.82	2.07	-	-	2.53	2.76	-
15th	-	-	-	-	-	-	-	-	-	-	-
16th	1.59	1.81	1.94	2.43	1.79	2.13	-	-	2.64	2.49	-
17th	1.57	1.75	1.70	2.96	1.72	2.06	-	-	-	-	2.26
18th	-	-	-	-	-	-	-	-	-	-	-
19th	-	-	-	-	-	-	-	-	-	-	-
20th	-	-	-	-	-	-	-	-	-	-	-
21st	-	-	-	-	-	-	-	-	-	-	-
22nd	-	-	-	-	-	-	-	-	-	-	-
23rd	1.80	1.73	1.75	2.30	1.75	2.01	-	-	-	-	-
24th	1.56	1.75	1.77	2.27	1.76	2.01	-	-	1.45	-	-
25th	1.66	1.76	1.71	2.27	1.74	2.01	-	-	2.45	-	-
26th	1.60	1.72	-	2.26	1.74	2.02	2.98	2.51	2.29	-	-
27th	1.60	1.72	1.79	2.26	1.72	2.00	3.00	2.47	2.25	-	-
28th	1.60	1.73	1.74	2.21	1.73	1.99	3.01	2.52	2.23	-	-
29th	1.64	1.76	1.76	2.25	1.77	2.03	3.08	2.62	2.29	-	-
30th	1.59	1.77	1.72	2.24	1.75	2.02	3.05	2.63	2.25	-	-
31st	1.55	-	-	-	-	-	-	-	2.93	-	-

Dissolved Oxygen (DO) (mg/l)

Date	Sample Locations										
	Inlet	W1	W2	W3	W4	W5	W6	W7	V1	V5	V10
1st	-	-	-	-	-	-	-	-	-	-	-
2nd	-	-	-	-	-	-	-	-	-	-	-
3rd	10.1	9.8	10.3	10.5	10.8	11.2	-	-	14.4	-	-
4th	9.8	10.2	10.6	10.1	11.0	11.1	-	-	13.2	-	-
5th	-	-	-	-	-	-	-	-	-	-	-
6th	-	-	-	-	-	-	-	-	-	-	-
7th	9.8	10.0	10.1	11.0	10.6	11.2	-	-	11.6	-	-
8th	-	-	-	-	-	-	-	-	-	-	-
9th	-	-	-	-	-	-	-	-	-	-	-
10th	10.8	10.4	9.8	11.1	10.6	9.9	-	-	11.8	-	-
11th	10.4	10.2	11.0	10.7	10.2	10.3	-	-	12.4	-	-
12th	11.6	11.4	11.3	11.2	12.4	11.8	-	-	13.8	12.4	
13th	-	-	-	-	-	-	-	-	-	-	-
14th	7.4	8.9	8.6	8.9	8.8	9.2	-	-	12.6	11.7	
15th	-	-	-	-	-	-	-	-	-	-	-
16th	12.4	11.8	11.6	12.7	11.8	11.2	-	-	12.4	12.6	
17th	10.8	11.2	10.9	11.2	10.6	10.7	-	-			10.4
18th	-	-	-	-	-	-	-	-	-	-	-
19th	-	-	-	-	-	-	-	-	-	-	-
20th	-	-	-	-	-	-	-	-	-	-	-
21st	-	-	-	-	-	-	-	-	-	-	-
22nd	-	-	-	-	-	-	-	-	-	-	-
23rd	6.8	6.7	7.4	8.5	7.2	8.1	-	-		-	-
24th	2.6	6.7	6.8	6.4	6.6	6.8	-	-	3.8	-	-
25th	8.1	8.8	8.2	9.0	8.9	9.6	-	-	10.0	-	-
26th	8.2	8.0	-	9.8	10.5	10.7	13.8	10.5	10.2	-	-
27th	7.2	9.2	9.3	9.5	10.2	10.3	11.6	9.7	10.0	-	-
28th	6.8	9.1	9.0	9.2	9.2	9.6	8.3	7.2	9.2	-	-
29th	9.2	11.5	11.8	12.2	12.0	11.5	10.8	15.1	12.0	-	-
30th	8.6	9.5	10.3	10.7	10.2	10.7	12.3	14.1	10.8	-	-
31st	-	-	-	-	-	-	-	-	-	-	-

Appendix B: Reed beds Photography.

Reed Bed 1

Reed Bed 2

Reed Bed 4

Reed Bed 3

Appendix C: Percentage of weir and valves openings.

Sample	Mon 14th	Tue 15th	Wed 16th	Thur 17th	Fri 18th	Sat 19th	Sun 20th
W1a	0	0	0	0	0	0	0
W1b	0	0	0	0	0	0	0
W2a	50	50	50	50	50	50	50
W2b	50	50	50	50	50	50	50
W3a	0	0	0	0	0	0	0
W3b	0	0	0	0	0	0	0
W4a	0	0	0	0	0	0	0
W4b	0	0	0	0	0	0	0
W5a	50	50	50	50	50	50	50
W5b	50	50	50	50	50	50	50
W6	25	25	25	25	25	25	25
W7	25	25	25	25	25	25	25
W8	0	0	0	0	0	0	0
W9	25	25	25	25	25	25	25
W10	0	0	0	0	0	0	0
W11	0	0	0	0	0	0	0
W12	0	0	0	0	0	0	0
W13	0	0	0	0	0	0	0
W14	0	0	0	0	0	0	0
W15	0	0	0	0	0	0	0
W16	0	0	0	0	0	0	0
W17	0	0	0	0	0	0	0
W18	0	0	0	0	0	0	0
W19	0	0	0	0	0	0	0
W20	0	0	0	0	0	0	0
W21	0	0	0	0	0	0	0
W22	0	0	0	0	0	0	0
W23	0	0	0	0	0	0	0
W24	0	0	0	0	0	0	0
W25	0	0	0	0	0	0	0
V1	0	0	0	0	0	0	0
V2	0	0	0	0	0	0	0
V3	0	0	0	0	0	0	0
V4	0	0	0	0	0	25	0
V5	0	0	0	0	0	0	0
V6	0	0	0	0	0	0	0
V7	0	0	0	0	0	25	0
V8	0	0	0	0	0	0	0
V9	0	0	0	0	0	0	0
V10	0	0	0	0	0	0	0
V11	0	0	0	0	0	0	0
V12	0	0	0	0	0	0	0
V13	0	0	0	0	0	0	0
V14	0	0	0	0	0	0	0
V15	0	0	0	0	0	0	0
V16	0	0	0	0	0	0	0

Oil Biodegradation

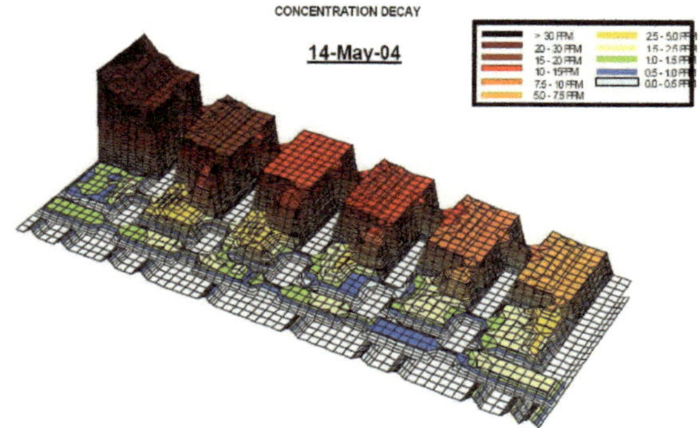

YOUR KNOWLEDGE HAS VALUE

- We will publish your bachelor's and
  master's thesis, essays and papers

- Your own eBook and book -
  sold worldwide in all relevant shops

- Earn money with each sale

Upload your text at www.GRIN.com
and publish for free